Archimedes

by Thomas Little Heath

ARCHIMEDES

BY

SIR THOMAS HEATH K.C.B., K.C.V.O., F.R.S.; SC.D., CAMB. HON. D.SC., OXFORD

[Greek: Dos moi pou sto, kai kino ten gen]

CONTENTS.

CHAPTER I.

ARCHIMEDES.

If the ordinary person were asked to say off-hand what he knew of Archimedes, he would probably, at the most, be able to quote one or other of the well-known stories about him: how, after discovering the solution of some problem in the bath, he was so overjoyed that he ran naked to his house, shouting [Greek: eureka, eureka] (or, as we might say, "I've got it, I've got it"); or how he said "Give me a place to stand on and I will move the earth"; or again how he was killed, at the capture of Syracuse in the Second Punic War, by a Roman soldier who resented being told to get away from a diagram drawn on the ground which he was studying.

And it is to be feared that few who are not experts in the history of mathematics have any acquaintance with the details of the original discoveries in mathematics of the greatest mathematician of antiquity, perhaps the greatest mathematical genius that the world has ever seen.

History and tradition know Archimedes almost exclusively as the inventor of a number of ingenious mechanical appliances, things which naturally appeal more to the popular imagination than the subtleties of pure mathematics.

Almost all that is told of Archimedes reaches us through the accounts by Polybius and Plutarch of the siege of Syracuse by Marcellus. He perished in the sack of that city in 212 B.C., and, as he was then an old man (perhaps 75 years old), he must have been born about 287 B.C. He was the son of Phidias, an astronomer, and was a friend and kinsman of King Hieron of Syracuse and his son Gelon. He spent some time at Alexandria studying with the successors of Euclid (Euclid who flourished about 300 B.C. was then no longer living). It was doubtless at Alexandria that he made the acquaintance of Conon of Samos, whom he admired as a mathematician and cherished as a friend, as well as of Eratosthenes; to the former, and to the latter during his early period he was in the habit of communicating his discoveries before their publication. It was also probably in Egypt that he invented the water-screw known by his name, the immediate purpose being the drawing of water for irrigating fields.

After his return to Syracuse he lived a life entirely devoted to mathematical research. Incidentally he became famous through his clever mechanical inventions. These things were, however, in his case the "diversions of geometry at play," and he attached no importance to them. In the words of Plutarch, "he possessed so lofty a spirit, so profound a soul, and such a wealth of scientific knowledge that, although these inventions had won for him the renown of more than human sagacity, yet he would not consent to leave behind him any written work on such subjects, but, regarding as ignoble and sordid the business of mechanics and every sort of art which is directed to practical utility, he placed his whole ambition in those speculations in the beauty and subtlety of which there is no admixture of the common needs of life".

During the siege of Syracuse Archimedes contrived all sorts of engines against the Roman besiegers. There were catapults so ingeniously constructed as to be equally serviceable at long or short range, and machines for discharging showers of missiles through holes made in the walls. Other machines consisted of long movable poles projecting beyond the walls; some of these dropped heavy weights upon the enemy's ships and on the constructions which they called sambuca, from their resemblance to a musical instrument of that name, and which consisted of a protected ladder with one end resting on two quinqueremes lashed together side by side as base, and capable of being raised by a windlass; others were fitted with an iron hand or a beak like that of a crane, which grappled the prows of ships, then lifted them into the air and let them fall again. Marcellus is said to have derided his own engineers and artificers with the words, "Shall we not make an end of fighting with this geometrical Briareus who uses our ships like cups to ladle water from the sea, drives our sambuca off ignominiously with cudgel-blows, and, by the multitude of missiles that he hurls at us all at once, outdoes the hundred-handed giants of mythology?" But the exhortation had no effect, the Romans being in such abject terror that, "if they did but see a piece of rope or wood projecting above the wall they would cry 'there it is,' declaring that Archimedes was setting some engine in motion against them, and would turn their backs and run away, insomuch that Marcellus desisted from all fighting and assault, putting all his hope in a long siege".

Archimedes died, as he had lived, absorbed in mathematical contemplation. The accounts of the circumstances of his death differ in some details. Plutarch gives more than one version in the following passage: "Marcellus was most of all afflicted at the death of Archimedes, for, as fate would have it, he was intent on working out some problem with a diagram, and, his mind and his eyes being alike fixed on his investigation, he never noticed the incursion of the Romans nor the capture of the city. And when a soldier came up to him suddenly and bade him follow to Marcellus, he refused to do so until he had worked out his problem to a demonstration; whereat the soldier was so enraged that he drew his sword and slew him. Others say that the Roman ran up to him with a drawn sword, threatening to kill him; and, when Archimedes saw him, he begged him earnestly to wait a little while in order that he might not leave his problem incomplete and unsolved, but the other took no notice and killed him. Again, there is a third account to the effect that, as he was carrying to Marcellus some of his mathematical instruments, sundials, spheres, and angles adjusted to the apparent size of the sun to the sight, some soldiers met him and, being under the impression that he carried gold in the vessel, killed him." The most picturesque version of the story is that which represents him as saying to a Roman soldier who came too close, "Stand away, fellow, from my diagram," whereat the man was so enraged that he killed him.

Archimedes is said to have requested his friends and relatives to place upon his tomb a representation of a cylinder circumscribing a sphere within it, together with an inscription giving the ratio (3/2) which the cylinder bears to the sphere; from which we may infer that he himself regarded the discovery of this ratio as his greatest achievement. Cicero, when quaestor in Sicily, found the tomb in a neglected state and restored it. In modern times not the slightest trace of it has been found.

Beyond the above particulars of the life of Archimedes, we have nothing but a number of stories which, if perhaps not literally accurate, yet help us to a conception of the personality of the man which we would not willingly have altered. Thus, in illustration of his entire preoccupation by his abstract studies, we are told that he would forget all about his food and such necessities of life, and would be drawing geometrical figures in the ashes of the fire, or, when anointing himself, in the oil on his body. Of the same kind is the story mentioned above, that, having discovered while in a bath the solution of the question referred to him by Hieron as to whether a certain crown supposed to have been made of gold did not in fact contain a certain proportion of silver, he ran naked through the street to his home shouting [Greek: eureka, eureka].

It was in connexion with his discovery of the solution of the problem To move a given weight by a given force that Archimedes uttered the famous saying, "Give me a place to stand on, and I can move the earth" ([Greek: dos moi pou sto kai kino ten gen], or in his broad Doric, as one version has it, [Greek: pa bo kai kino tan gan]). Plutarch represents him as declaring to Hieron that any given weight could be moved by a given force, and boasting, in reliance on the cogency of his demonstration, that, if he were given another earth, he would cross over to it and move this one. "And when Hieron was struck with amazement and asked him to reduce the problem to practice and to show him some great weight moved by a small force, he fixed on a ship of burden with three masts from the king's arsenal which had only been drawn up by the great labour of many men; and loading her with many passengers and a full freight, sitting himself the while afar off, with no great effort but quietly setting in motion with his hand a compound pulley, he drew the ship towards him smoothly and safely as if she were moving through the sea." Hieron, we are told elsewhere, was so much astonished that he declared that, from that day forth, Archimedes's word was to be accepted on every subject! Another version of the story describes the machine used as a helix; this term must be supposed to refer to a screw in the shape of a cylindrical helix turned by a handle and acting on a cog-wheel with oblique teeth fitting on the screw.

Another invention was that of a sphere constructed so as to imitate the motions of the sun, the moon, and the five planets in the heavens. Cicero actually saw this contrivance, and he gives a description of it, stating that it represented the periods of the moon and the apparent motion of the sun with such accuracy that it would even (over a short period) show the eclipses of the sun and moon. It may have been moved by water, for Pappus speaks in one place of "those who understand the making of spheres and produce a model of the heavens by means of the regular circular motion of water". In any case it is certain that Archimedes was much occupied with astronomy. Livy calls him "unicus spectator caeli siderumque". Hipparchus says, "From these observations it is clear that the differences in the years are altogether small, but, as to the solstices, I almost think that both I and Archimedes have erred to the extent of a quarter of a day both in observation and in the deduction therefrom." It appears, therefore, that Archimedes had considered the question of the length of the year. Macrobius says that he discovered the distances of the planets. Archimedes himself describes in the Sandreckoner the apparatus by which he measured the apparent diameter of the sun, i.e. the angle subtended by it at the eye.

The story that he set the Roman ships on fire by an arrangement of burning-glasses or concave mirrors is not found in any authority earlier than Lucian (second century A.D.); but there is no improbability in the idea that he discovered some form of burning-mirror, e.g. a paraboloid of revolution, which would reflect to one point all rays falling on its concave surface in a direction parallel to its axis.

CHAPTER II.

GREEK GEOMETRY TO ARCHIMEDES.

In order to enable the reader to arrive at a correct understanding of the place of Archimedes and of the significance of his work it is necessary to pass in review the course of development of Greek geometry from its first beginnings down to the time of Euclid and Archimedes.

Greek authors from Herodotus downwards agree in saying that geometry was invented by the Egyptians and that it came into Greece from Egypt. One account says:--

"Geometry is said by many to have been invented among the Egyptians, its origin being due to the measurement of plots of land. This was necessary there because of the rising of the Nile, which obliterated the boundaries appertaining to separate owners. Nor is it marvellous that the discovery of this and the other sciences should have arisen from such an occasion, since everything which moves in the sense of development will advance from the imperfect to the perfect. From sense-perception to reasoning, and from reasoning to understanding, is a natural transition. Just as among the Phoenicians, through commerce and exchange, an accurate knowledge of numbers was originated, so also among the Egyptians geometry was invented for the reason above stated.

"Thales first went to Egypt and thence introduced this study into Greece."

But it is clear that the geometry of the Egyptians was almost entirely practical and did not go beyond the requirements of the land-surveyor, farmer or merchant. They did indeed know, as far back as 2000 B.C., that in a triangle which has its sides proportional to 3, 4, 5 the angle contained by the two smaller sides is a right angle, and they used such a triangle as a practical means of drawing right angles. They had formulae, more or less inaccurate, for certain measurements, e.g. for the areas of certain triangles, parallel-trapezia, and circles. They had, further, in their construction of pyramids, to use the notion of similar right-angled triangles; they even had a name, se-qet, for the ratio of the half of the side of the base to the height, that is, for what we should call the co-tangent of the angle of slope. But not a single general theorem in geometry can be traced to the Egyptians. Their knowledge that the triangle (3, 4, 5) is right angled is far from implying any knowledge of the general proposition (Eucl. I., 47) known by the name of Pythagoras. The science of geometry, in fact, remained to be discovered; and this required the genius for pure speculation which the Greeks possessed in the largest measure among all the nations of the world.

Thales, who had travelled in Egypt and there learnt what the priests could teach him on the subject, introduced geometry into Greece. Almost the whole of Greek science and philosophy begins with Thales. His date was about 624-547 B.C. First of the Ionian philosophers, and declared one of the Seven Wise Men in 582-581, he shone in all fields, as astronomer, mathematician, engineer, statesman and man of business. In astronomy he predicted the solar eclipse of 28 May, 585, discovered the inequality of the four astronomical seasons, and counselled the use of the Little Bear instead of the Great Bear as a means of finding the pole. In geometry the following theorems are attributed to him--and their character shows how the Greeks had to begin at the very beginning of the theory--(1) that a circle is bisected by any diameter (Eucl. I., Def. 17), (2) that the angles at the base of an isosceles triangle are equal (Eucl. I., 5), (3) that, if two straight lines cut one another, the vertically opposite angles are equal (Eucl. I., 15), (4) that, if two triangles have two angles and one side respectively equal, the triangles are equal in all respects (Eucl. I., 26). He is said (5) to have been the first to inscribe a right-angled triangle in a circle: which must mean that he was the first to discover that the angle in a semicircle is a right angle. He also solved two problems in practical geometry: (1) he showed how to measure the distance from the land of a ship at sea (for this he is said to have used the proposition numbered (4) above), and (2) he measured the heights of pyramids by means of the shadow thrown on the ground (this implies the use of similar triangles in the way that the Egyptians had used them in the construction of pyramids).

After Thales come the Pythagoreans. We are told that the Pythagoreans were the first to use the term [Greek: mathemata] (literally "subjects of instruction") in the specialised sense of "mathematics"; they, too, first advanced mathematics as a study pursued for its own sake and made it a part of a liberal education. Pythagoras, son of Mnesarchus, was born in Samos about 572 B.C., and died at a great age (75 or 80) at Metapontum. His interests were as various as those of Thales; his travels, all undertaken in pursuit of knowledge, were probably even more extended. Like Thales, and perhaps at his suggestion, he visited Egypt and studied there for a long period (22 years, some say).

It is difficult to disentangle from the body of Pythagorean doctrines the portions which are due to Pythagoras himself because of the habit which the members of the school had of attributing everything to the Master ([Greek: autos epha], ipse dixit). In astronomy two things at least may safely be attributed to him; he held that the earth is spherical in shape, and he recognised that the sun, moon and planets have an independent motion of their own in a direction contrary to that of the daily rotation; he seems, however, to have adhered to the geocentric view of the universe, and it was his successors who evolved the theory that the earth does not remain at the centre but revolves, like the other planets and the sun and moon, about the "central fire". Perhaps his most remarkable discovery was the dependence of the musical intervals on the lengths of vibrating strings, the proportion for the octave being 2 : 1, for the fifth 3 : 2 and for the fourth 4 : 3. In arithmetic he was the first to expound the theory of means and of proportion as applied to commensurable quantities. He laid the foundation of the theory of numbers by considering the properties of numbers as such, namely, prime numbers, odd and even numbers, etc. By means of figured numbers, square, oblong, triangular, etc. (represented by dots arranged in the form of the various figures) he showed the connexion between numbers and geometry. In view of all these properties of numbers, we can easily understand how the Pythagoreans came to "liken all things to numbers" and to find in the principles of numbers the principles of all things ("all things are numbers").

We come now to Pythagoras's achievements in geometry. There is a story that, when he came home from Egypt and tried to found a school at Samos, he found the Samians indifferent, so that he had to take special measures to ensure that his geometry might not perish with him. Going to the gymnasium, he sought out a well-favoured youth who seemed likely to suit his purpose, and was withal poor, and bribed him to learn geometry by promising him sixpence for every proposition that he mastered. Very soon the youth got fascinated by the subject for its own sake, and Pythagoras rightly judged that he would gladly go on without the sixpence. He hinted, therefore, that he himself was poor and must try to earn his living instead of doing mathematics; whereupon the youth, rather than give up the study, volunteered to pay sixpence to Pythagoras for each proposition.

In geometry Pythagoras set himself to lay the foundations of the subject, beginning with certain important definitions and investigating the fundamental principles. Of propositions attributed to him the most famous is, of course, the theorem that in a right-angled triangle the square on the hypotenuse is equal to the sum of the squares on the sides about the right angle (Eucl. I., 47); and, seeing that Greek tradition universally credits him with the proof of this theorem, we prefer to believe that tradition is right. This is to some extent confirmed by another tradition that Pythagoras discovered a general formula for finding two numbers such that the sum of their squares is a square number. This depends on the theory of the gnomon, which at first had an arithmetical signification corresponding to the geometrical use of it in Euclid, Book II. A figure in the shape of a gnomon put round two sides of a square makes it into a larger square. Now consider the number 1 represented by a dot. Round this place three other dots so that the four dots form a square $(1 + 3 = 2^2)$. Round the four dots (on two adjacent sides of the square) place five dots at regular and equal distances, and we have another square $(1 + 3 + 5 = 3^2)$; and so on. The successive odd numbers 1, 3, 5 ... were called gnomons, and the general formula is

$$1 + 3 + 5 + ... + (2n - 1) = n^2.$$

Add the next odd number, i.e. $2n + 1$, and we have $n^2 + (2n + 1) = (n + 1)^2$. In order, then, to get two square numbers such that their sum is a square we have only to see that $2n + 1$ is a square. Suppose that $2n + 1 = m^2$; then $n =$

1/2(m^2 - 1), and we have {1/2(m^2 - 1)}^2 + m^2 = {1/2(m^2 + 1)}^2, where m is any odd number; and this is the general formula attributed to Pythagoras.

Proclus also attributes to Pythagoras the theory of proportionals and the construction of the five "cosmic figures," the five regular solids.

One of the said solids, the dodecahedron, has twelve pentagonal faces, and the construction of a regular pentagon involves the cutting of a straight line "in extreme and mean ratio" (Eucl. II., 11, and VI., 30), which is a particular case of the method known as the application of areas. How much of this was due to Pythagoras himself we do not know; but the whole method was at all events fully worked out by the Pythagoreans and proved one of the most powerful of geometrical methods. The most elementary case appears in Euclid, I., 44, 45, where it is shown how to apply to a given straight line as base a parallelogram having a given angle (say a rectangle) and equal in area to any rectilineal figure; this construction is the geometrical equivalent of arithmetical division. The general case is that in which the parallelogram, though applied to the straight line, overlaps it or falls short of it in such a way that the part of the parallelogram which extends beyond, or falls short of, the parallelogram of the same angle and breadth on the given straight line itself (exactly) as base is similar to another given parallelogram (Eucl. VI., 28, 29). This is the geometrical equivalent of the most general form of quadratic equation $ax [+-] mx^2 = C$, so far as it has real roots; while the condition that the roots may be real was also worked out (= Eucl. VI., 27). It is important to note that this method of application of areas was directly used by Apollonius of Perga in formulating the fundamental properties of the three conic sections, which properties correspond to the equations of the conics in Cartesian co-ordinates; and the names given by Apollonius (for the first time) to the respective conics are taken from the theory, parabola ([Greek: parabole]) meaning "application" (i.e. in this case the parallelogram is applied to the straight line exactly), hyperbola([Greek: hyperbole]), "exceeding" (i.e. in this case the parallelogram exceeds or overlaps the straight line), ellipse ([Greek: elleipsis]), "falling short" (i.e. the parallelogram falls short of the straight line).

Another problem solved by the Pythagoreans is that of drawing a rectilineal figure equal in area to one given rectilineal figure and similar to another. Plutarch mentions a doubt as to whether it was this problem or the proposition of Euclid I., 47, on the strength of which Pythagoras was said to have sacrificed an ox.

The main particular applications of the theorem of the square on the hypotenuse (e.g. those in Euclid, Book II.) were also Pythagorean; the construction of a square equal to a given rectangle (Eucl. II., 14) is one of them and corresponds to the solution of the pure quadratic equation $x^2 = ab$.

The Pythagoreans proved the theorem that the sum of the angles of any triangle is equal to two right angles (Eucl. I., 32).

Speaking generally, we may say that the Pythagorean geometry covered the bulk of the subject-matter of Books I., II., IV., and VI. of Euclid (with the qualification, as regards Book VI., that the Pythagorean theory of proportion applied only to commensurable magnitudes). Our information about the origin of the propositions of Euclid, Book III., is not so complete; but it is certain that the most important of them were well known to Hippocrates of Chios (who flourished in the second half of the fifth century, and lived perhaps from about 470 to 400 B.C.), whence we conclude that the main propositions of Book III. were also included in the Pythagorean geometry.

Lastly, the Pythagoreans discovered the existence of incommensurable lines, or of irrationals. This was, doubtless, first discovered with reference to the diagonal of a square which is incommensurable with the side, being in the ratio to it of [root]2 to 1. The Pythagorean proof of this particular case survives in Aristotle and in a proposition interpolated in Euclid's Book X.; it is by a reductio ad absurdumproving that, if the diagonal is commensurable with the side, the same number must be both odd and even. This discovery of the incommensurable was bound to cause geometers a great shock, because it showed that the theory of proportion invented by Pythagoras was not of universal application, and therefore that propositions proved by means of it were not really established. Hence the stories that the discovery of the irrational was for a time kept secret, and that the first person who divulged it perished by shipwreck. The fatal flaw thus revealed in the body of geometry was not removed till Eudoxus (408-355 B.C.) discovered the great theory of proportion (expounded in Euclid's Book V.), which is applicable to incommensurable as well as to commensurable magnitudes.

By the time of Hippocrates of Chios the scope of Greek geometry was no longer even limited to the Elements; certain special problems were also attacked which were beyond the power of the geometry of the straight line and circle, and which were destined to play a great part in determining the direction taken by Greek geometry in its highest flights. The main problems in question were three: (1) the doubling of the cube, (2) the trisection of any angle, (3) the squaring of the circle; and from the time of Hippocrates onwards the investigation of these problems proceeded pari passu with the completion of the body of the Elements.

Hippocrates himself is an example of the concurrent study of the two departments. On the one hand, he was the first of the Greeks who is known to have compiled a book of Elements. This book, we may be sure, contained in particular the most important propositions about the circle included in Euclid, Book III. But a much more important proposition is attributed to Hippocrates; he is said to have been the first to prove that circles are to one another as the squares on their diameters (= Eucl. XII., 2), with the deduction that similar segments of circles are to one another as the squares on their bases. These propositions were used by him in his tract on the squaring of lunes, which was intended to lead up to the squaring of the circle. The latter problem is one which must have exercised practical geometers from time immemorial. Anaxagoras for instance (about 500-428 B.C.) is said to have worked at the problem while in prison. The essential portions of Hippocrates's tract are preserved in a passage of Simplicius (on Aristotle's Physics), which contains substantial fragments from Eudemus's History of

Geometry. Hippocrates showed how to square three particular lunes of different forms, and then, lastly, he squared the sum of a certain circle and a certain lune. Unfortunately, however, the last-mentioned lune was not one of those which can be squared, and so the attempt to square the circle in this way failed after all.

Hippocrates also attacked the problem of doubling the cube. There are two versions of the origin of this famous problem. According to one of them, an old tragic poet represented Minos as having been dissatisfied with the size of a tomb erected for his son Glaucus, and having told the architect to make it double the size, retaining, however, the cubical form. According to the other, the Delians, suffering from a pestilence, were told by the oracle to double a certain cubical altar as a means of staying the plague. Hippocrates did not, indeed, solve the problem, but he succeeded in reducing it to another, namely, the problem of finding two mean proportionals in continued proportion between two given straight lines, i.e. finding x, y such that $a : x = x : y = y : b$, where a, b are the two given straight lines. It is easy to see that, if $a : x = x : y = y : b$, then $b/a = (x/a)^3$, and, as a particular case, if $b = 2a$, $x^3 = 2a^3$, so that the side of the cube which is double of the cube of side a is found.

The problem of doubling the cube was henceforth tried exclusively in the form of the problem of the two mean proportionals. Two significant early solutions are on record.

(1) Archytas of Tarentum (who flourished in first half of fourth century B.C.) found the two mean proportionals by a very striking construction in three dimensions, which shows that solid geometry, in the hands of Archytas at least, was already well advanced. The construction was usually called mechanical, which it no doubt was in form, though in reality it was in the highest degree theoretical. It consisted in determining a point in space as the intersection of three surfaces: (a) a cylinder, (b) a cone, (c) an "anchor-ring" with internal radius = 0. (2) Menaechmus, a pupil of Eudoxus, and a contemporary of Plato, found the two mean proportionals by means of conic

sections, in two ways, ([alpha]) by the intersection of two parabolas, the equations of which in Cartesian co-ordinates would be $x^2 = ay$, $y^2 = bx$, and ([beta]) by the intersection of a parabola and a rectangular hyperbola, the corresponding equations being $x^2 = ay$, and $xy = ab$ respectively. It would appear that it was in the effort to solve this problem that Menaechmus discovered the conic sections, which are called, in an epigram by Eratosthenes, "the triads of Menaechmus".

The trisection of an angle was effected by means of a curve discovered by Hippias of Elis, the sophist, a contemporary of Hippocrates as well as of Democritus and Socrates (470-399 B.C.). The curve was called the quadratrix because it also served (in the hands, as we are told, of Dinostratus, brother of Menaechmus, and of Nicomedes) for squaring the circle. It was theoretically constructed as the locus of the point of intersection of two straight lines moving at uniform speeds and in the same time, one motion being angular and the other rectilinear. Suppose OA, OB are two radii of a circle at right angles to one another. Tangents to the circle at A and B, meeting at C, form with the two radii the square OACB. The radius OA is made to move uniformly about O, the centre, so as to describe the angle AOB in a certain time. Simultaneously AC moves parallel to itself at uniform speed such that A just describes the line AO in the same length of time. The intersection of the moving radius and AC in their various positions traces out the quadratrix.

The rest of the geometry which concerns us was mostly the work of a few men, Democritus of Abdera, Theodorus of Cyrene (the mathematical teacher of Plato), Theaetetus, Eudoxus, and Euclid. The actual writers of Elements of whom we hear were the following. Leon, a little younger than Eudoxus (408-355 B.C.), was the author of a collection of propositions more numerous and more serviceable than those collected by Hippocrates. Theudius of Magnesia, a contemporary of Menaechmus and Dinostratus, "put together the elements admirably, making many partial or limited propositions more general". Theudius's book was no doubt the geometrical text-book of the Academy and that used by Aristotle.

Theodorus of Cyrene and Theaetetus generalised the theory of irrationals, and we may safely conclude that a great part of the substance of Euclid's Book X. (on irrationals) was due to Theaetetus. Theaetetus also wrote on the five regular solids (the tetrahedron, cube, octahedron, dodecahedron, and icosahedron), and Euclid was therefore no doubt equally indebted to Theaetetus for the contents of his Book XIII. In the matter of Book XII. Eudoxus was the pioneer. These facts are confirmed by the remark of Proclus that Euclid, in compiling his Elements, collected many of the theorems of Eudoxus, perfected many others by Theaetetus, and brought to irrefragable demonstration the propositions which had only been somewhat loosely proved by his predecessors.

Eudoxus (about 408-355 B.C.) was perhaps the greatest of all Archimedes's predecessors, and it is his achievements, especially the discovery of the method of exhaustion, which interest us in connexion with Archimedes.

In astronomy Eudoxus is famous for the beautiful theory of concentric spheres which he invented to explain the apparent motions of the planets, and, particularly, their apparent stationary points and retrogradations. The theory applied also to the sun and moon, for which Eudoxus required only three spheres in each case. He represented the motion of each planet as compounded of the rotations of four interconnected spheres about diameters, all of which pass through the centre of the earth. The outermost sphere represents the daily rotation, the second a motion along the zodiac circle or ecliptic; the poles of the third sphere, about which that sphere revolves, are fixed at two opposite points on the zodiac circle, and are carried round in the motion of the second sphere; and on the surface of the third sphere the poles of the fourth sphere are fixed; the fourth sphere, revolving about the diameter joining its two poles, carries the planet which is fixed at a point on its equator. The poles and the speeds and directions of rotation are so chosen that the planet actually describes a hippopede, or horse-fetter, as it was called (i.e. a figure of eight), which lies along and is longitudinally bisected by the zodiac circle, and is carried round that circle. As a tour de force of geometrical imagination it would be difficult to parallel this hypothesis.

In geometry Eudoxus discovered the great theory of proportion, applicable to incommensurable as well as commensurable magnitudes, which is expounded in Euclid, Book V., and which still holds its own and will do so for all time. He also solved the problem of the two mean proportionals by means of certain curves, the nature of which, in the absence of any description of them in our sources, can only be conjectured.

Last of all, and most important for our purpose, is his use of the famous method of exhaustion for the measurement of the areas of curves and the volumes of solids. The example of this method which will be most familiar to the reader is the proof in Euclid XII., 2, of the theorem that the areas of circles are to one another as the squares on their diameters. The proof in this and in all cases depends on a lemma which forms Prop. 1 of Euclid's Book X. to the effect that, if there are two unequal magnitudes of the same kind and from the greater you subtract not less than its half, then from the remainder not less than its half, and so on continually, you will at length have remaining a magnitude less than the lesser of the two magnitudes set out, however small it is. Archimedes says that the theorem of Euclid XII., 2, was proved by means of a certain lemma to the effect that, if we have two unequal magnitudes (i.e. lines, surfaces, or solids respectively), the greater exceeds the lesser by such a magnitude as is capable, if added continually to itself, of exceeding any magnitude of the same kind as the original magnitudes. This assumption is known as the Axiom or Postulate of Archimedes, though, as he states, it was assumed before his time by those who used the method of exhaustion. It is in reality used in Euclid's lemma (Eucl. X., 1) on which Euclid XII., 2, depends, and only differs in statement from Def. 4 of Euclid, Book V., which is no doubt due to Eudoxus.

The method of exhaustion was not discovered all at once; we find traces of gropings after such a method before it was actually evolved. It was perhaps Antiphon, the sophist, of Athens, a contemporary of Socrates (470-399 B.C.), who took the first step. He inscribed a square (or, according to another account, an equilateral triangle) in a circle, then bisected the arcs subtended by the sides, and so inscribed a polygon of double the number of sides; he then repeated the process, and maintained that, by continuing it, we should at last arrive at a polygon with sides so small as to make the polygon coincident with the circle. Though this was formally incorrect, it nevertheless contained the germ of the method of exhaustion.

Hippocrates, as we have seen, is said to have proved the theorem that circles are to one another as the squares on their diameters, and it is difficult to see how he could have done this except by some form, or anticipation, of the method. There is, however, no doubt about the part taken by Eudoxus; he not only based the method on rigorous demonstration by means of the lemma or lemmas aforesaid, but he actually applied the method to find the volumes (1) of any pyramid, (2) of the cone, proving (1) that any pyramid is one third part of the prism which has the same base and equal height, and (2) that any cone is one third part of the cylinder which has the same base and equal height. Archimedes, however, tells us the remarkable fact that these two theorems were first discovered by Democritus (who flourished towards the end of the fifth century B.C.), though he was not able to prove them (which no doubt means, not that he gave no sort of proof, but that he was not able to establish the propositions by the rigorous method of Eudoxus). Archimedes adds that we must give no small share of the credit for these theorems to Democritus; and this is another testimony to the marvellous powers, in mathematics as well as in other subjects, of the great man who, in the words of Aristotle, "seems to have thought of everything". We know from other sources that Democritus wrote on irrationals; he is also said to have discussed the question of two parallel sections of a cone (which were evidently supposed to be indefinitely close together), asking whether we are to regard them as unequal or equal: "for if they are unequal they will make the cone irregular as having many indentations, like steps, and unevennesses, but, if they are equal, the cone will appear to have the property of the cylinder and to be made up of equal, not unequal, circles, which is very absurd". This explanation shows that Democritus was already close on the track of infinitesimals.

Archimedes says further that the theorem that spheres are in the triplicate ratio of their diameters was proved by means of the same lemma. The proofs of the propositions about the volumes of pyramids, cones and spheres are, of course, contained in Euclid, Book XII. (Props. 3-7 Cor., 10, 16-18 respectively).

It is no doubt desirable to illustrate Eudoxus's method by one example. We will take one of the simplest, the proposition (Eucl. XII., 10) about the cone. Given ABCD, the circular base of the cylinder which has the same base as the cone and equal height, we inscribe the square ABCD; we then bisect the arcs subtended by the sides, and draw the regular inscribed polygon of eight sides, then similarly we draw the regular inscribed polygon of sixteen sides, and so on. We erect on each regular polygon the prism which has the polygon for base, thereby obtaining successive prisms inscribed in the cylinder, and of the same height with it. Each time we double the number of sides in the base of the prism we take away more than half of the volume by which the cylinder exceeds the prism (since we take away more than half of the excess of the area of the circular base over that of the inscribed polygon, as in Euclid XII., 2). Suppose now that V is the volume of the cone, C that of the cylinder. We have to prove that C = 3V. If C is not equal to 3V, it is either greater or less than 3V.

Suppose (1) that C > 3V, and that C = 3V + E. Continue the construction of prisms inscribed in the cylinder until the parts of the cylinder left over outside the final prism (of volume P) are together less than E.

Then C - P < E. But C - 3V = E; Therefore P > 3V.

But it has been proved in earlier propositions that P is equal to three times the pyramid with the same base as the prism and equal height.

Therefore that pyramid is greater than V, the volume of the cone: which is impossible, since the cone encloses the pyramid.

Therefore C is not greater than 3V.

Next (2) suppose that C < 3V, so that, inversely,

V > 1/3 C.

This time we inscribe successive pyramids in the cone until we arrive at a pyramid such that the portions of the cone left over outside it are together less than the excess of V over 1/3 C. It follows that the pyramid is greater than 1/3 C. Hence the prism on the same base as the pyramid and inscribed in the cylinder (which prism is three times the pyramid) is greater than C: which is impossible, since the prism is enclosed by the cylinder, and is therefore less than it.

Therefore V is not greater than 1/3 C, or C is not less than 3V.

Accordingly C, being neither greater nor less than 3V, must be equal to it; that is, V = 1/3 C.

It only remains to add that Archimedes is fully acquainted with the main properties of the conic sections. These had already been proved in earlier treatises, which Archimedes refers to as the "Elements of Conics". We know of two such treatises, (1) Euclid's four Books on Conics, (2) a work by one Aristaeus called "Solid Loci," probably a treatise on conics regarded as loci. Both these treatises are lost; the former was, of course, superseded by Apollonius's great work on Conics in eight Books.

CHAPTER III.

THE WORKS OF ARCHIMEDES.

The range of Archimedes's writings will be gathered from the list of his various treatises. An extraordinarily large proportion of their contents represents entirely new discoveries of his own. He was no compiler or writer of text-books, and in this respect he differs from Euclid and Apollonius, whose work largely consisted in systematising and generalising the methods used and the results obtained by earlier geometers. There is in Archimedes no mere working-up of existing material; his objective is always something new, some definite addition to the sum of knowledge. Confirmation of this is found in the introductory letters prefixed to most of his treatises. In them we see the directness, simplicity and humanity of the man. There is full and generous recognition of the work of predecessors and contemporaries; his estimate of the relation of his own discoveries to theirs is obviously just and free from any shade of egoism. His manner is to state what particular discoveries made by his predecessors had suggested to him the possibility of extending them in new directions; thus he says that, in connexion with the efforts of earlier geometers to square the circle, it occurred to him that no one had tried to square a parabolic segment; he accordingly attempted the problem and finally solved it. Similarly he describes his discoveries about the volumes and surfaces of spheres and cylinders as supplementing the theorems of Eudoxus about the pyramid, the cone and the cylinder. He does not hesitate to say that certain problems baffled him for a long time; in one place he positively insists, for the purpose of pointing a moral, on specifying two propositions which he had enunciated but which on further investigation proved to be wrong.

The ordinary MSS. of the Greek text of Archimedes give his works in the following order:--

1. On the Sphere and Cylinder (two books). 2. Measurement of a Circle. 3. On Conoids and Spheroids. 4. On Spirals. 5. On Plane Equilibriums (two books). 6. The Sandreckoner. 7. Quadrature of a Parabola.

A most important addition to this list has been made in recent years through an extraordinary piece of good fortune. In 1906 J. L. Heiberg, the most recent editor of the text of Archimedes, discovered a palimpsest of mathematical content in the "Jerusalemic Library" of one Papadopoulos Kerameus at Constantinople. This proved to contain writings of Archimedes copied in a good hand of the tenth century. An attempt had been made (fortunately with only partial success) to wash out the old writing, and then the parchment was used again to write a Euchologion upon. However, on most of the leaves the earlier writing remains more or less legible. The important fact about the MS. is that it contains, besides substantial portions of the treatises previously known, (1) a considerable portion of the work, in two books, On Floating Bodies, which was formerly supposed to have been lost in Greek and only to have survived in the translation by Wilhelm of Morbeke, and (2) most precious of all, the greater part of the book called The Method, treating of Mechanical Problems and addressed to Eratosthenes. The important treatise so happily recovered is now included in Heiberg's new (second) edition of the Greek text of Archimedes (Teubner, 1910-15), and some account of it will be given in the next chapter.

The order in which the treatises appear in the MSS. was not the order of composition; but from the various prefaces and from internal evidence generally we are able to establish the following as being approximately the chronological sequence:--

1. On Plane Equilibriums, I. 2. Quadrature of a Parabola. 3. On Plane Equilibriums, II. 4. The Method. 5. On the Sphere and Cylinder, I, II. 6. On Spirals. 7. On Conoids and Spheroids. 8. On Floating Bodies, I, II. 9. Measurement of a Circle. 10. The Sandreckoner.

In addition to the above we have a collection of geometrical propositions which has reached us through the Arabic with the title "Liber assumptorum Archimedis". They were not written by Archimedes in their present form, but were probably collected by some later Greek writer for the purpose of illustrating some ancient work. It is, however, quite likely that some of the propositions, which are remarkably elegant, were of Archimedean origin, notably those concerning the geometrical figures made with three and four semicircles respectively and called (from their shape) (1) the shoemaker's knifeand (2) the Salinon or salt-cellar, and another theorem which bears on the trisection of an angle.

An interesting fact which we now know from Arabian sources is that the formula for the area of any triangle in terms of its sides which we write in the form

$$[\Delta] = [root]\{s(s - a)(s - b)(s - c)\},$$

and which was supposed to be Heron's because Heron gives the geometrical proof of it, was really due to Archimedes.

Archimedes is further credited with the authorship of the famous Cattle-Problem enunciated in a Greek epigram edited by Lessing in 1773. According to its heading the problem was communicated by Archimedes to the mathematicians at Alexandria in a letter to Eratosthenes; and a scholium to Plato's Charmides speaks of the problem "called by Archimedes the Cattle-Problem". It is an extraordinarily difficult problem in indeterminate analysis, the solution of which involves enormous figures.

Of lost works of Archimedes the following can be identified:--

1. Investigations relating to polyhedra are referred to by Pappus, who, after speaking of the five regular solids, gives a description of thirteen other polyhedra discovered by Archimedes which are semi-regular, being contained by polygons equilateral and equiangular but not similar. One at least of these semi-regular solids was, however, already known to Plato.

2. A book of arithmetical content entitled Principles dealt, as we learn from Archimedes himself, with the naming of numbers, and expounded a system of expressing large numbers which could not be written in the ordinary Greek notation. In setting out the same system in the Sandreckoner (see Chapter V. below), Archimedes explains that he does so for the benefit of those who had not seen the earlier work.

3. On Balances (or perhaps levers). Pappus says that in this work Archimedes

proved that "greater circles overpower lesser circles when they rotate about the same centre".

4. A book On Centres of Gravity is alluded to by Simplicius. It is not, however, certain that this and the last-mentioned work were separate treatises, Possibly Book I. On Plane Equilibriums may have been part of a larger work (called perhaps Elements of Mechanics), and On Balances may have been an alternative title. The title On Centres of Gravity may be a loose way of referring to the same treatise.

5. Catoptrica, an optical work from which Theon of Alexandria quotes a remark about refraction.

6. On Sphere-making, a mechanical work on the construction of a sphere to represent the motions of the heavenly bodies (cf. pp. 5-6 above).

Arabian writers attribute yet further works to Archimedes, (1) On the circle, (2) On a heptagon in a circle, (3) On circles touching one another, (4) On parallel lines, (5) On triangles, (6) On the properties of right-angled triangles, (7) a book of Data; but we have no confirmation of these statements.

CHAPTER IV.

GEOMETRY IN ARCHIMEDES.

The famous French geometer, Chasles, drew an instructive distinction between the predominant features of the geometry of the two great successors of Euclid, namely, Archimedes and Apollonius of Perga (the "great geometer," and author of the classical treatise on Conics). The works of these two men may, says Chasles, be regarded as the origin and basis of two great inquiries which seem to share between them the domain of geometry. Apollonius is concerned with the Geometry of Forms and Situations, while in Archimedes we find the Geometry of Measurements, dealing with the quadrature of curvilinear plane figures and with the quadrature and cubature of curved surfaces, investigations which gave birth to the calculus of the infinite conceived and brought to perfection by Kepler, Cavalieri, Fermat, Leibniz and Newton.

In geometry Archimedes stands, as it were, on the shoulders of Eudoxus in that he applied the method of exhaustion to new and more difficult cases of quadrature and cubature. Further, in his use of the method he introduced an interesting variation of the procedure as we know it from Euclid. Euclid (and presumably Eudoxus also) only used inscribedfigures, "exhausting" the figure to be measured, and had to invert the second half of the reductio ad absurdum to enable approximation from below (so to speak) to be applied in that case also. Archimedes, on the other hand, approximates from above as well as from below; he approaches the area or volume to be measured by taking closer and closer circumscribed figures, as well as inscribed, and thereby compressing, as it were, the inscribed and circumscribed figure into one, so that they ultimately coincide with one another and with the figure to be measured. But he follows the cautious method to which the Greeks always adhered; he never says that a given curve or surface is the limiting form of the inscribed or circumscribed figure; all that he asserts is that we can approach the curve or surface as nearly as we please.

The deductive form of proof by the method of exhaustion is apt to obscure not only the way in which the results were arrived at but also the real character of the procedure followed. What Archimedes actually does in certain cases is to perform what are seen, when the analytical equivalents are set down, to be real integrations; this remark applies to his investigation of the areas of a parabolic segment and a spiral respectively, the surface and volume respectively of a sphere and a segment of a sphere, and the volume of any segments of the solids of revolution of the second degree. The result is, as a rule, only obtained after a long series of preliminary propositions, all of which are links in a chain of argument elaborately forged for the one purpose. The method suggests the tactics of some master of strategy who foresees everything, eliminates everything not immediately conducive to the execution of his plan, masters every position in its order, and then suddenly (when the very elaboration of the scheme has almost obscured, in the mind of the onlooker, its ultimate object) strikes the final blow. Thus we read in Archimedes proposition after proposition the bearing of which is not immediately obvious but which we find infallibly used later on; and we are led on by such easy stages that the difficulty of the original problem, as presented at the outset, is scarcely appreciated. As Plutarch says, "It is not possible to find in geometry more difficult and troublesome questions, or more simple and lucid explanations". But it is decidedly a rhetorical exaggeration when Plutarch goes on to say that we are deceived by the easiness of the successive steps into the belief that any one could have discovered them for himself. On the contrary, the studied simplicity and the perfect finish of the treatises involve at the same time an element of mystery. Although each step depends upon the preceding ones, we are left in the dark as to how they were suggested to Archimedes. There is, in fact, much truth in a remark of Wallis to the effect that he seems "as it were of set purpose to have covered up the traces of his investigation as if he had grudged posterity the secret of his method of inquiry while he wished to extort from them assent to his results".

A partial exception is now furnished by the Method; for here we have (as it were) a lifting of the veil and a glimpse of the interior of Archimedes's workshop. He tells us how he discovered certain theorems in quadrature and cubature, and he is at the same time careful to insist on the difference between (1) the means which may serve to suggest the truth of theorems, although not furnishing scientific proofs of them, and (2) the rigorous demonstrations of them by approved geometrical methods which must follow before they can be finally accepted as established.

Writing to Eratosthenes he says: "Seeing in you, as I say, an earnest student, a man of considerable eminence in philosophy and an admirer of mathematical inquiry when it comes your way, I have thought fit to write out for you and explain in detail in the same book the peculiarity of a certain method, which, when you see it, will put you in possession of a means whereby you can investigate some of the problems of mathematics by mechanics. This procedure is, I am persuaded, no less useful for the proofs of the actual theorems as well. For certain things which first became clear to me by a mechanical method had afterwards to be demonstrated by geometry, because their investigation by the said method did not furnish an actual demonstration. But it is of course easier, when we have previously acquired by the method some knowledge of the questions, to supply the proof than it is to find the proof without any previous knowledge. This is a reason why, in the case of the theorems the proof of which Eudoxus was the first to discover, namely, that the cone is a third part of the cylinder, and the pyramid a third part of the prism, having the same base and equal height, we should give no small share of the credit to Democritus, who was the first to assert this truth with regard to the said figures, though he did not prove it. I am myself in the position of having made the discovery of the theorem now to be published in the same way as I made my earlier discoveries; and I thought it desirable now to write out and publish the method, partly because I have already spoken of it and I do not want to be thought to have uttered vain words, but partly also because I am persuaded that it will be of no little service to mathematics; for I apprehend that some, either of my contemporaries or of my successors, will, by means of the method when once established, be able to discover other theorems in addition, which have not occurred to me.

"First then I will set out the very first theorem which became known to me by means of mechanics, namely, that Any segment of a section of a right-angled cone [i.e. a parabola] is four-thirds of the triangle which has the same base and equal height; and after this I will give each of the other theorems investigated by the same method. Then, at the end of the book, I will give the geometrical proofs of the propositions."

The following description will, I hope, give an idea of the general features of the mechanical method employed by Archimedes. Suppose that X is the plane or solid figure the area or content of which is to be found. The method in the simplest case is to weigh infinitesimal elements of X against the corresponding elements of another figure, B say, being such a figure that its area or content and the position of its centre of gravity are already known.

The diameter or axis of the figure X being drawn, the infinitesimal elements taken are parallel sections of X in general, but not always, at right angles to the axis or diameter, so that the centres of gravity of all the sections lie at one point or other of the axis or diameter and their weights can therefore be taken as acting at the several points of the diameter or axis. In the case of a plane figure the infinitesimal sections are spoken of as parallel straight lines and in the case of a solid figure as parallel planes, and the aggregate of the infinite number of sections is said to make up the whole figure X. (Although the sections are so spoken of as straight lines or planes, they are really indefinitely narrow plane strips or indefinitely thin laminae respectively.) The diameter or axis is produced in the direction away from the figure to be measured, and the diameter or axis as produced is imagined to be the bar or lever of a balance. The object is now to apply all the separate elements of X at one point on the lever, while the corresponding elements of the known figure B operate at different points, namely, where they actually arein the first instance. Archimedes contrives, therefore, to move the elements of X away from their original position and to concentrate them at one point on the lever, such that each of the elements balances, about the point of suspension of the lever, the corresponding element of B acting at its centre of gravity. The elements of X and B respectively balance about the point of suspension in accordance with the property of the lever that the weights are inversely proportional to the distances from the fulcrum or point of suspension. Now the centre of gravity of B as a whole is known, and it may then be supposed to act as one mass at its centre of gravity. (Archimedes assumes as known that the sum of the "moments," as we call them, of all the elements of the figure B, acting severally at the points where they actually are, is equal to the moment of the whole figure applied as one mass at one point, its centre of gravity.) Moreover all the elements of X are concentrated at the one fixed point on the bar or lever. If this fixed point is H, and G is the centre of gravity of the figure B, while C is the point of suspension,

X : B = CG : CH.

Thus the area or content of X is found.

Conversely, the method can be used to find the centre of gravity of X when its area or volume is known beforehand. In this case the elements of X, and X itself, have to be applied where they are, and the elements of the known figure or figures have to be applied at the one fixed point H on the other side of C, and since X, B and CH are known, the proportion

B : X = CG : CH

determines CG, where G is the centre of gravity of X.

The mechanical method is used for finding (1) the area of any parabolic segment, (2) the volume of a sphere and a spheroid, (3) the volume of a segment of a sphere and the volume of a right segment of each of the three conicoids of revolution, (4) the centre of gravity (a) of a hemisphere, (b) of any segment of a sphere, (c) of any right segment of a spheroid and a paraboloid of revolution, and (d) of a half-cylinder, or, in other words, of a semicircle.

Archimedes then proceeds to find the volumes of two solid figures, which are the special subject of the treatise. The solids arise as follows:--

(1) Given a cylinder inscribed in a rectangular parallelepiped on a square base in such a way that the two bases of the cylinder are circles inscribed in the opposite square faces, suppose a plane drawn through one side of the square containing one base of the cylinder and through the parallel diameter of the opposite base of the cylinder. The plane cuts off a solid with a surface resembling that of a horse's hoof. Archimedes proves that the volume of the solid so cut off is one sixth part of the volume of the parallelepiped.

(2) A cylinder is inscribed in a cube in such a way that the bases of the cylinder are circles inscribed in two opposite square faces. Another cylinder is inscribed which is similarly related to another pair of opposite faces. The two cylinders include between them a solid with all its angles rounded off; and

Archimedes proves that the volume of this solid is two-thirds of that of the cube.

Having proved these facts by the mechanical method. Archimedes concluded the treatise with a rigorous geometrical proof of both propositions by the method of exhaustion. The MS. is unfortunately somewhat mutilated at the end, so that a certain amount of restoration is necessary.

I shall now attempt to give a short account of the other treatises of Archimedes in the order in which they appear in the editions. The first is--

On the Sphere and Cylinder.

Book I. begins with a preface addressed to Dositheus (a pupil of Conon), which reminds him that on a former occasion he had communicated to him the treatise proving that any segment of a "section of a right-angled cone" (i.e. a parabola) is four-thirds of the triangle with the same base and height, and adds that he is now sending the proofs of certain theorems which he has since discovered, and which seem to him to be worthy of comparison with Eudoxus's propositions about the volumes of a pyramid and a cone. The theorems are (1) that the surface of a sphere is equal to four times its greatest circle (i.e. what we call a "great circle" of the sphere); (2) that the surface of any segment of a sphere is equal to a circle with radius equal to the straight line drawn from the vertex of the segment to a point on the circle which is the base of the segment; (3) that, if we have a cylinder circumscribed to a sphere and with height equal to the diameter, then (a) the volume of the cylinder is 1-1/2 times that of the sphere and (b) the surface of the cylinder, including its bases, is 1-1/2 times the surface of the sphere.

Next come a few definitions, followed by certain Assumptions, two of which are well known, namely:--

1. Of all lines which have the same extremities the straight line is the least (this has been made the basis of an alternative definition of a straight line).

2. Of unequal lines, unequal surfaces and unequal solids the greater exceeds the less by such a magnitude as, when (continually) added to itself, can be made to exceed any assigned magnitude among those which are comparable [with it and] with one another (i.e. are of the same kind). This is the Postulate of Archimedes.

He also assumes that, of pairs of lines (including broken lines) and pairs of surfaces, concave in the same direction and bounded by the same extremities, the outer is greater than the inner. These assumptions are fundamental to his investigation, which proceeds throughout by means of figures inscribed and circumscribed to the curved lines or surfaces that have to be measured.

After some preliminary propositions Archimedes finds (Props. 13, 14) the area of the surfaces (1) of a right cylinder, (2) of a right cone. Then, after quoting certain Euclidean propositions about cones and cylinders, he passes to the main business of the book, the measurement of the volume and surface of a sphere and a segment of a sphere. By circumscribing and inscribing to a great circle a regular polygon of an even number of sides and making it revolve about a diameter connecting two opposite angular points he obtains solids of revolution greater and less respectively than the sphere. In a series of propositions he finds expressions for (a) the surfaces, (b) the volumes, of the figures so inscribed and circumscribed to the sphere. Next he proves (Prop. 32) that, if the inscribed and circumscribed polygons which, by their revolution, generate the figures are similar, the surfaces of the figures are in the duplicate ratio, and their volumes in the triplicate ratio, of their sides. Then he proves that the surfaces and volumes of the inscribed and circumscribed figures respectively are less and greater than the surface and volume respectively to which the main propositions declare the surface and volume of the sphere to be equal (Props. 25, 27, 30, 31 Cor.). He has now all the material for applying the method of exhaustion and so proves the main propositions about the surface and volume of the sphere. The rest of the book applies the same procedure to a segment of the sphere. Surfaces of revolution are inscribed and circumscribed to a segment less than a hemisphere, and the theorem about the surface of the segment is finally proved in Prop. 42. Prop. 43 deduces the surface of a segment greater than a hemisphere. Prop. 44 gives the volume of the sector of the sphere which includes any segment.

Book II begins with the problem of finding a sphere equal in volume to a given cone or cylinder; this requires the solution of the problem of the two mean proportionals, which is accordingly assumed. Prop. 2 deduces, by means of 1., 44, an expression for the volume of a segment of a sphere, and Props. 3, 4 solve the important problems of cutting a given sphere by a plane so that (a) the surfaces, (b) the volumes, of the segments may have to one another a given ratio. The solution of the second problem (Prop. 4) is difficult. Archimedes reduces it to the problem of dividing a straight line AB into two parts at a point M such that

MB : (a given length) = (a given area) : AM^2.

The solution of this problem with a determination of the limits of possibility are given in a fragment by Archimedes, discovered and preserved for us by Eutocius in his commentary on the book; they are effected by means of the points of intersection of two conics, a parabola and a rectangular hyperbola. Three problems of construction follow, the first two of which are to construct a segment of a sphere similar to one given segment, and having (a) its volume, (b) its surface, equal to that of another given segment of a sphere. The last two propositions are interesting. Prop. 8 proves that, if V, V' be the volumes, and S, S' the surfaces, of two segments into which a sphere is divided by a plane, V and S belonging to the greater segment, then

$S^2 : S'^2 > V : V' > S^{(3/2)} : S'^{(3/2)}$.

Prop. 9 proves that, of all segments of spheres which have equal surfaces, the hemisphere is the greatest in volume.

The Measurement of a Circle.

This treatise, in the form in which it has come down to us, contains only three propositions; the second, being an easy deduction from Props. 1 and 3, is out of place in so far as it uses the result of Prop. 3.

In Prop. 1 Archimedes inscribes and circumscribes to a circle a series of successive regular polygons, beginning with a square, and continually doubling the number of sides; he then proves in the orthodox manner by the method of exhaustion that the area of the circle is equal to that of a right-angled triangle, in which the perpendicular is equal to the radius, and the base equal to the circumference, of the circle. Prop. 3 is the famous proposition in which Archimedes finds by sheer calculation upper and lower arithmetical limits to the ratio of the circumference of a circle to its diameter, or what we call [pi]; the result obtained is 3-1/7> [pi] > 3-10/71. Archimedes inscribes and circumscribes successive regular polygons, beginning with hexagons, and doubling the number of sides continually, until he arrives at inscribed and circumscribed regular polygons with 96 sides; seeing then that the length of the circumference of the circle is intermediate between the perimeters of the two polygons, he calculates the two perimeters in terms of the diameter of the circle. His calculation is based on two close approximations (an upper and a lower) to the value of [root]3, that being the cotangent of the angle of 30 deg., from which he begins to work. He assumes as known that 265/153 < [root]3 < 1351/780. In the text, as we have it, only the results of the steps in the calculation are given, but they involve the finding of approximations to the square roots of several large numbers: thus 1172-1/8 is given as the approximate value of [root](1373943-33/64), 3013-3/4 as that of [root](9082321) and 1838-9/11 as that of [root](3380929). In this way Archimedes arrives at 14688/(4673-1/2) as the ratio of the perimeter of the circumscribed polygon of 96 sides to the diameter of the circle; this is the figure which he rounds up into 3-1/7. The corresponding figure for the inscribed polygon is 6336/(2017-1/4), which, he says, is > 3-10/71. This example shows how little the Greeks were embarrassed in arithmetical calculations by their alphabetical system of numerals.

On Conoids and Spheroids.

The preface addressed to Dositheus shows, as we may also infer from internal evidence, that the whole of this book also was original. Archimedes first explains what his conoids and spheroids are, and then, after each description, states the main results which it is the aim of the treatise to prove. The conoids are two. The first is the right-angled conoid, a name adapted from the old name ("section of a right-angled cone") for a parabola; this conoid is therefore a paraboloid of revolution. The second is the obtuse-angled conoid, which is a hyperboloid of revolution described by the revolution of a hyperbola (a "section of an obtuse-angled cone") about its transverse axis. The spheroids are two, being the solids of revolution described by the revolution of an ellipse (a "section of an acute-angled cone") about (1) its major axis and (2) its minor axis; the first is called the "oblong" (or oblate) spheroid, the second the "flat" (or prolate) spheroid. As the volumes of oblique segments of conoids and spheroids are afterwards found in terms of the volume of the conical figure with the base of the segment as base and the vertex of the segment as vertex, and as the said base is thus an elliptic section of an oblique circular cone, Archimedes calls the conical figure with an elliptic base a "segment of a cone" as distinct from a "cone".

As usual, a series of preliminary propositions is required. Archimedes first sums, in geometrical form, certain series, including the arithmetical progression, a, 2a, 3a, ... na, and the series formed by the squares of these terms (in other words the series 1^2, 2^2, 3^2, ... n^2); these summations are required for the final addition of an indefinite number of elements of each figure, which amounts to an integration. Next come two properties of conics (Prop. 3), then the determination by the method of exhaustion of the area of an ellipse (Prop. 4). Three propositions follow, the first two of which (Props. 7, 8) show that the conical figure above referred to is really a segment of an oblique circular cone; this is done by actually finding the circular sections. Prop. 9 gives a similar proof that each elliptic section of a conoid or spheroid is a section of a certain oblique circular cylinder (with axis parallel to the axis of the segment of the conoid or spheroid cut off by the said elliptic section).

Props. 11-18 show the nature of the various sections which cut off segments of each conoid and spheroid and which are circles or ellipses according as the section is perpendicular or obliquely inclined to the axis of the solid; they include also certain properties of tangent planes, etc.

The real business of the treatise begins with Props. 19, 20; here it is shown how, by drawing many plane sections equidistant from one another and all parallel to the base of the segment of the solid, and describing cylinders (in general oblique) through each plane section with generators parallel to the axis of the segment and terminated by the contiguous sections on either side, we can make figures circumscribed and inscribed to the segment, made up of segments of cylinders with parallel faces and presenting the appearance of the steps of a staircase. Adding the elements of the inscribed and circumscribed figures respectively and using the method of exhaustion, Archimedes finds the volumes of the respective segments of the solids in the approved manner (Props. 21, 22 for the paraboloid, Props. 25, 26 for the hyperboloid, and Props. 27-30 for the spheroids). The results are stated in this form: (1) Any segment of a paraboloid of revolution is half as large again as the cone or segment of a cone which has the same base and axis; (2) Any segment of a hyperboloid of revolution or of a spheroid is to the cone or segment of a cone with the same base and axis in the ratio of AD + 3CA to AD + 2CA in the case of the hyperboloid, and of 3CA - AD to 2CA - AD in the case of the spheroid, where C is the centre, A the vertex of the segment, and AD the axis of the segment (supposed in the case of the spheroid to be not greater than half the spheroid).

On Spirals.

The preface addressed to Dositheus is of some length and contains, first, a tribute to the memory of Conon, and next a summary of the theorems about the sphere and the conoids and spheroids included in the above two treatises. Archimedes then passes to the spiral which, he says, presents another sort of problem, having nothing in common with the foregoing. After a definition of the spiral he enunciates the main propositions about it which are to be proved in the treatise. The spiral (now known as the Spiral of Archimedes) is defined as the locus of a point starting from a given point (called the "origin") on a given straight line and moving along the straight line at uniform speed, while the line itself revolves at uniform speed about the origin as a fixed point. Props. 1-11 are preliminary, the last two amounting to the summation of certain series required for the final addition of an indefinite number of element-areas, which again amounts to integration, in order to find the area of the figure cut off between any portion of the curve and the two radii vectores drawn to its extremities. Props. 13-20 are interesting and difficult propositions establishing the properties of tangents to the spiral. Props. 21-23 show how to inscribe and circumscribe to any portion of the spiral figures consisting of a multitude of elements which are narrow sectors of circles with the origin as centre; the area of the spiral is intermediate between the areas of the inscribed and circumscribed figures, and by the usual method of exhaustion Archimedes finds the areas required. Prop. 24 gives the area of the first complete turn of the spiral (= $1/3[pi](2[pi]a)^2$, where the spiral is r = a[theta]), and of any portion of it up to OP where P is any point on the first turn. Props. 25, 26 deal similarly with the second turn of the spiral and with the area subtended by any arc (not being greater than a complete turn) on any turn. Prop. 27 proves the interesting property that, if R1 be the area of the first turn of the spiral bounded by the initial line, R2 the area of the ring added by the second complete turn, R3 the area of the ring added by the third turn, and so on, then R3 = 2R2, R4 = 3R2, R5 = 4R2, and so on to Rn = (n - 1)R2, while R2, = 6R1.

Quadrature of the Parabola.

The title of this work seems originally to have been On the Section of a Right-angled Cone and to have been changed after the time of Apollonius, who was the first to call a parabola by that name. The preface addressed to Dositheus was evidently the first communication from Archimedes to him after the death of Conon. It begins with a feeling allusion to his lost friend, to whom the treatise was originally to have been sent. It is in this preface that Archimedes alludes to the lemma used by earlier geometers as the basis of the method of exhaustion (the Postulate of Archimedes, or the theorem of Euclid X., 1). He mentions as having been proved by means of it (1) the theorems that the areas of circles are to one another in the duplicate ratio of their diameters, and that the volumes of spheres are in the triplicate ratio of their diameters, and (2) the propositions proved by Eudoxus about the volumes of a cone and a pyramid. No one, he says, so far as he is aware, has yet tried to square the segment bounded by a straight line and a section of a right-angled cone (a parabola); but he has succeeded in proving, by means of the same lemma, that the parabolic segment is equal to four-thirds of the triangle on the same base and of equal height, and he sends the proofs, first as "investigated" by means of mechanics and secondly as "demonstrated" by geometry. The phraseology shows that here, as in the Method, Archimedes regarded the mechanical investigation as furnishing evidence rather than proof of the truth of the proposition, pure geometry alone furnishing the absolute proof required.

The mechanical proof with the necessary preliminary propositions about the parabola (some of which are merely quoted, while two, evidently original, are proved, Props. 4, 5) extends down to Prop. 17; the geometrical proof with other auxiliary propositions completes the book (Props. 18-24). The mechanical proof recalls that of the Method in some respects, but is more elaborate in that the elements of the area of the parabola to be measured are not straight lines but narrow strips. The figures inscribed and circumscribed to the segment are made up of such narrow strips and have a saw-like edge; all the elements are trapezia except two, which are triangles, one in each figure. Each trapezium (or triangle) is weighed where it is against another area hung at a fixed point of an assumed lever; thus the whole of the inscribed and circumscribed figures respectively are weighed against the sum of an indefinite number of areas all suspended from one point on the lever. The result is obtained by a real integration, confirmed as usual by a proof by the method of exhaustion.

The geometrical proof proceeds thus. Drawing in the segment the inscribed triangle with the same base and height as the segment, Archimedes next inscribes triangles in precisely the same way in each of the segments left over, and proves that the sum of the two new triangles is 1/4 of the original inscribed triangle. Again, drawing triangles inscribed in the same way in the four segments left over, he proves that their sum is 1/4 of the sum of the preceding pair of triangles and therefore $(1/4)^2$ of the original inscribed triangle. Proceeding thus, we have a series of areas exhausting the parabolic segment. Their sum, if we denote the first inscribed triangle by [Delta], is

[Delta]$\{1 + 1/4 + (1/4)^2 + (1/4)^3 + ...\}$

Archimedes proves geometrically in Prop. 23 that the sum of this infinite series is 4/3[Delta], and then confirms by reductio ad absurdum the equality of the area of the parabolic segment to this area.

CHAPTER V.

THE SANDRECKONER.

The Sandreckoner deserves a place by itself. It is not mathematically very important; but it is an arithmetical curiosity which illustrates the versatility and genius of Archimedes, and it contains some precious details of the history of Greek astronomy which, coming from such a source and at first hand, possess unique authority. We will begin with the astronomical data. They are contained in the preface addressed to King Gelon of Syracuse, which begins as follows:--

"There are some, King Gelon, who think that the number of the sand is infinite in multitude; and I mean by the sand not only that which exists about Syracuse and the rest of Sicily but also that which is found in every region whether inhabited or uninhabited. Again, there are some who, without regarding it as infinite, yet think that no number has been named which is great enough to exceed its multitude. And it is clear that they who hold this view, if they imagined a mass made up of sand in other respects as large as the mass of the earth, including in it all the seas and the hollows of the earth filled up to a height equal to that of the highest of the mountains, would be many times further still from recognising that any number could be expressed which exceeded the multitude of the sand so taken. But I will try to show you, by means of geometrical proofs which you will be able to follow, that, of the numbers named by me and given in the work which I sent to Zeuxippus, some exceed not only the number of the mass of sand equal in size to the earth filled up in the way described, but also that of a mass equal in size to the universe.

"Now you are aware that 'universe' is the name given by most astronomers to the sphere the centre of which is the centre of the earth, while the radius is equal to the straight line between the centre of the sun and the centre of the earth. This is the common account, as you have heard from astronomers. But Aristarchus of Samos brought out a book consisting of some hypotheses, in which the premises lead to the conclusion that the universe is many times greater than that now so called. His hypotheses are that the fixed stars and the sun remain unmoved, that the earth revolves about the sun in the circumference of a circle, the sun lying in the centre of the orbit, and that the sphere of the fixed stars, situated about the same centre as the sun, is so great that the circle in which he supposes the earth to revolve bears such a ratio to the distance of the fixed stars as the centre of the sphere bears to its surface."

Here then is absolute and practically contemporary evidence that the Greeks, in the person of Aristarchus of Samos (about 310-230 B.C.), had anticipated Copernicus.

By the last words quoted Aristarchus only meant to say that the size of the earth is negligible in comparison with the immensity of the universe. This, however, does not suit Archimedes's purpose, because he has to assume a definite size, however large, for the universe. Consequently he takes a liberty with Aristarchus. He says that the centre (a mathematical point) can have no ratio whatever to the surface of the sphere, and that we must therefore take Aristarchus to mean that the size of the earth is to that of the so-called "universe" as the size of the so-called "universe" is to that of the real universe in the new sense.

Next, he has to assume certain dimensions for the earth, the moon and the sun, and to estimate the angle subtended at the centre of the earth by the sun's diameter; and in each case he has to exaggerate the probable figures so as to be on the safe side. While therefore (he says) some have tried to prove that the perimeter of the earth is 300,000 stadia (Eratosthenes, his contemporary, made it 252,000 stadia, say 24,662 miles, giving a diameter of about 7,850 miles), he will assume it to be ten times as great or 3,000,000 stadia. The diameter of the earth, he continues, is greater than that of the moon and that of the sun is greater than that of the earth. Of the diameter of the sun he observes that Eudoxus had declared it to be nine times that of the moon, and his own father, Phidias, had made it twelve times, while Aristarchus had tried to prove that the diameter of the sun is greater than eighteen times but less than twenty times the diameter of the moon (this was in the treatise of Aristarchus On the Sizes and Distances of the Sun and Moon, which is still extant, and is an admirable piece of geometry, proving rigorously, on the basis of certain assumptions, the result stated). Archimedes again intends to be on the safe side, so he takes the diameter of the sun to be thirty times that of the moon and not greater. Lastly, he says that Aristarchus discovered that the diameter of the sun appeared to be about 1/720th part of the zodiac circle, i.e. to subtend an angle of about half a degree; and he describes a simple instrument by which he himself found that the angle subtended by the diameter of the sun at the time when it had just risen was less than 1/164th part and greater than 1/200th part of a right angle. Taking this as the size of the angle subtended at the eye of the observer on the surface of the earth, he works out, by an interesting geometrical proposition, the size of the angle subtended at the centre of the earth, which he finds to be > 1/203rd part of a right angle. Consequently the diameter of the sun is greater than the side of a regular polygon of 812 sides inscribed in a great circle of the so-called "universe," and a fortiori greater than the side of a regular chiliagon (polygon of 1000 sides) inscribed in that circle.

On these assumptions, and seeing that the perimeter of a regular chiliagon (as of any other regular polygon of more than six sides) inscribed in a circle is more than 3 times the length of the diameter of the circle, it easily follows that, while the diameter of the earth is less than 1,000,000 stadia, the diameter of the so-called "universe" is less than 10,000 times the diameter of the earth, and therefore less than 10,000,000,000 stadia.

Lastly, Archimedes assumes that a quantity of sand not greater than a poppy-seed contains not more than 10,000 grains, and that the diameter of a poppy-seed is not less than 1/40th of a dactylus (while a stadium is less than 10,000 dactyli).

Archimedes is now ready to work out his calculation, but for the inadequacy of the alphabetic system of numerals to express such large numbers as are required. He, therefore, develops his remarkable terminology for expressing large numbers.

The Greek has names for all numbers up to a myriad (10,000); there was, therefore, no difficulty in expressing with the ordinary numerals all numbers up to a myriad myriads (100,000,000). Let us, says Archimedes, call all these numbers numbers of the first order. Let the second order of numbers begin with 100,000,000, and end with $100,000,000^2$. Let $100,000,000^2$ be the first number of the third order, and let this extend to $100,000,000^3$; and so on, to the myriad-myriadth order, beginning with $100,000,000^{(99,999,999)}$ and ending with $100,000,000^{(100,000,000)}$, which for brevity we will call P. Let all the numbers of all the orders up to P form the first period, and let the first order of the second period begin with P and end with 100,000,000 P; let the second order begin with this, the third order with $100,000,000^2$ P, and so on up to the 100,000,000th order of the second period, ending with $1,000,000,000^{(100,000,000)}$ P or P^2. The first order of the third period begins with P^2, and the orders proceed as before. Continuing the series of periods and orders of each period, we finally arrive at the 100,000,000th period ending with $P^{(100,000,000)}$. The prodigious extent of this scheme is seen when it is considered that the last number of the first period would now be represented by 1 followed by 800,000,000 ciphers, while the last number of the 100,000,000th period would require 100,000,000 times as many ciphers, i.e. 80,000 million million ciphers.

As a matter of fact, Archimedes does not need, in order to express the "number of the sand," to go beyond the eighth order of the first period. The orders of the first period begin respectively with 1, 10^8, 10^{16}, 10^{24}, ... $(10^8)^{\wedge}(99,999,999)$; and we can express all the numbers required in powers of 10.

Since the diameter of a poppy-seed is not less than 1/40th of a dactylus, and spheres are to one another in the triplicate ratio of their diameters, a sphere of diameter 1 dactylus is not greater than 64,000 poppy-seeds, and, therefore, contains not more than 64,000 X 10,000 grains of sand, and a fortiori not more than 1,000,000,000, or 10^9 grains of sand. Archimedes multiplies the diameter of the sphere continually by 100, and states the corresponding number of grains of sand. A sphere of diameter 10,000 dactyli and a fortiori of one stadium contains less than 10^{21} grains; and proceeding in this way to spheres of diameter 100 stadia, 10,000 stadia and so on, he arrives at the number of grains of sand in a sphere of diameter 10,000,000,000 stadia, which is the size of the so-called universe; the corresponding number of grains of sand is 10^{51}. The diameter of the real universe being 10,000 times that of the so-called universe, the final number of grains of sand in the real universe is found to be 10^{63}, which in Archimedes's terminology is a myriad-myriad units of the eighth order of numbers.

CHAPTER VI.

MECHANICS.

It is said that Archytas was the first to treat mechanics in a systematic way by the aid of mathematical principles; but no trace survives of any such work by him. In practical mechanics he is said to have constructed a mechanical dove which would fly, and also a rattle to amuse children and "keep them from breaking things about the house" (so says Aristotle, adding "for it is impossible for children to keep still").

In the Aristotelian Mechanica we find a remark on the marvel of a great weight being moved by a small force, and the problems discussed bring in the lever in various forms as a means of doing this. We are told also that practically all movements in mechanics reduce to the lever and the principle of the lever (that the weight and the force are in inverse proportion to the distances from the point of suspension or fulcrum of the points at which they act, it being assumed that they act in directions perpendicular to the lever). But the lever is merely "referred to the circle"; the force which acts at the greater distance from the fulcrum is said to move a weight more easily because it describes a greater circle.

There is, therefore, no proof here. It was reserved for Archimedes to prove the property of the lever or balance mathematically, on the basis of certain postulates precisely formulated and making no large demand on the faith of the learner. The treatise On Plane Equilibriums in two books is, as the title implies, a work on statics only; and, after the principle of the lever or balance has been established in Props. 6, 7 of Book I., the rest of the treatise is devoted to finding the centre of gravity of certain figures. There is no dynamics in the work and therefore no room for the parallelogram of velocities, which is given with a fairly adequate proof in the Aristotelian Mechanica.

Archimedes's postulates include assumptions to the following effect: (1) Equal weights at equal distances are in equilibrium, and equal weights at unequal distances are not in equilibrium, but the system in that case "inclines towards the weight which is at the greater distance," in other words, the action of the weight which is at the greater distance produces motion in the direction in which it acts; (2) and (3) If when weights are in equilibrium something is added to or subtracted from one of the weights, the system will "incline" towards the weight which is added to or the weight from which nothing is taken respectively; (4) and (5) If equal and similar figures be applied to one another so as to coincide throughout, their centres of gravity also coincide; if figures be unequal but similar, their centres of gravity are similarly situated with regard to the figures.

The main proposition, that two magnitudes balance at distances reciprocally proportional to the magnitudes, is proved first for commensurable and then for incommensurable magnitudes. Preliminary propositions have dealt with equal magnitudes disposed at equal distances on a straight line and odd or even in number, and have shown where the centre of gravity of the whole system lies. Take first the case of commensurable magnitudes. If A, B be the weights acting at E, D on the straight line ED respectively, and ED be divided at C so that A : B = DC : CE, Archimedes has to prove that the system is in equilibrium about C. He produces ED to K, so that DK = EC, and DE to L so that EL = CD; LK is then a straight line bisected at C. Again, let H be taken on LK such that LH = 2LE or 2CD, and it follows that the remainder HK = 2DK or 2EC. Since A, B are commensurable, so are EC, CD. Let x be a common measure of EC, CD. Take a weight w such that w is the same part of A that x is of LH. It follows that w is the same part of B that x is of HK. Archimedes now divides LH, HK into parts equal to x, and A B into parts equal to w, and places the w's at the middle points of the x's respectively. All the w's are then in equilibrium about C. But all the w's acting at the several points along LH are equivalent to A acting as a whole at the point E. Similarly the w's acting at the several points on HK are equivalent to B acting at D. Therefore A, B placed at E, D respectively balance about C.

Prop. 7 deduces by reductio ad absurdum the same result in the case where A, B are incommensurable. Prop. 8 shows how to find the centre of gravity of the remainder of a magnitude when the centre of gravity of the whole and of a part respectively are known. Props. 9-15 find the centres of gravity of a parallelogram, a triangle and a parallel-trapezium respectively.

Book II., in ten propositions, is entirely devoted to finding the centre of gravity of a parabolic segment, an elegant but difficult piece of geometrical work which is as usual confirmed by the method of exhaustion.

CHAPTER VII.

HYDROSTATICS.

The science of hydrostatics is, even more than that of statics, the original creation of Archimedes. In hydrostatics he seems to have had no predecessors. Only one of the facts proved in his work On Floating Bodies, in two books, is given with a sort of proof in Aristotle. This is the proposition that the surface of a fluid at rest is that of a sphere with its centre at the centre of the earth.

Archimedes founds his whole theory on two postulates, one of which comes at the beginning and the other after Prop. 7 of Book I. Postulate 1 is as follows:--

"Let us assume that a fluid has the property that, if its parts lie evenly and are continuous, the part which is less compressed is expelled by that which is more compressed, and each of its parts is compressed by the fluid above it perpendicularly, unless the fluid is shut up in something and compressed by something else."

Postulate 2 is: "Let us assume that any body which is borne upwards in water is carried along the perpendicular [to the surface] which passes through the centre of gravity of the body".

In Prop. 2 Archimedes proves that the surface of any fluid at rest is the surface of a sphere the centre of which is the centre of the earth. Props. 3-7 deal with the behaviour, when placed in fluids, of solids (1) just as heavy as the fluid, (2) lighter than the fluid, (3) heavier than the fluid. It is proved (Props. 5, 6) that, if the solid is lighter than the fluid, it will not be completely immersed but only so far that the weight of the solid will be equal to that of the fluid displaced, and, if it be forcibly immersed, the solid will be driven upwards by a force equal to the difference between the weight of the solid and that of the fluid displaced. If the solid is heavier than the fluid, it will, if placed in the fluid, descend to the bottom and, if weighed in the fluid, the solid will be lighter than its true weight by the weight of the fluid displaced (Prop. 7).

The last-mentioned theorem naturally connects itself with the story of the crown made for Hieron. It was suspected that this was not wholly of gold but contained an admixture of silver, and Hieron put to Archimedes the problem of determining the proportions in which the metals were mixed. It was the discovery of the solution of this problem when in the bath that made Archimedes run home naked, shouting [Greek: eureka, eureka]. One account of the solution makes Archimedes use the proposition last quoted; but on the whole it seems more likely that the actual discovery was made by a more elementary method described by Vitruvius. Observing, as he is said to have done, that, if he stepped into the bath when it was full, a volume of water was spilt equal to the volume of his body, he thought of applying the same idea to the case of the crown and measuring the volumes of water displaced respectively (1) by the crown itself, (2) by the same weight of pure gold, and (3) by the same weight of pure silver. This gives an easy means of solution. Suppose that the weight of the crown is W, and that it contains weights w1 and w2, of gold and silver respectively. Now experiment shows (1) that the crown itself displaces a certain volume of water, V say, (2) that a weight W of gold displaces a certain other volume of water, V1 say, and (3) that a weight W of silver displaces a volume V2.

From (2) it follows, by proportion, that a weight w1 of gold will displace w1/W . V1 of the fluid, and from (3) it follows that a weight w2 of silver displaces w2/W . V2 of the fluid.

Hence V = w1/W . V1 + w2/W . V2;

therefore WV = w1V1 + w2V2,

that is, (w1 + w2)V = w1V1 + w2V2,

so that $w_1/w_2 = (V_2 - V)/(V - V_1)$,

which gives the required ratio of the weights of gold and silver contained in the crown.

The last two propositions of Book I. investigate the case of a segment of a sphere floating in a fluid when the base of the segment is (1) entirely above and (2) entirely below the surface of the fluid; and it is shown that the segment will in either case be in equilibrium in the position in which the axis is vertical, the equilibrium being in the first case stable.

Book II. is a geometrical tour de force. Here, by the methods of pure geometry, Archimedes investigates the positions of rest and stability of a right segment of a paraboloid of revolution floating with its base upwards or downwards (but completely above or completely below the surface) for a number of cases differing (1) according to the relation between the length of the axis of the paraboloid and the principal parameter of the generating parabola, and (2) according to the specific gravity of the solid in relation to the fluid; where the position of rest and stability is such that the axis of the solid is not vertical, the angle at which it is inclined to the vertical is fully determined.

The idea of specific gravity appears all through, though this actual term is not used. Archimedes speaks of the solid being lighter or heavier than the fluid or equally heavy with it, or when a ratio has to be expressed, he speaks of a solid the weight of which (for an equal volume) has a certain ratio to that of the fluid.

BIBLIOGRAPHY.

The editio princeps of the works of Archimedes with the commentaries of Eutocius was brought out by Hervagius (Herwagen) at Basel in 1544. D. Rivault (Paris, 1615) gave the enunciations in Greek and the proofs in Latin somewhat retouched. The Arenarius (Sandreckoner) and the Dimensio circuli with Eutocius's commentary were edited with Latin translation and notes by Wallis in 1678 (Oxford). Torelli's monumental edition (Oxford, 1792) of the Greek text of the complete works and of the commentaries of Eutocius, with a new Latin translation, remained the standard text until recent years; it is now superseded by the definitive text with Latin translation of the complete works, Eutocius's commentaries, the fragments, scholia, etc., edited by Heiberg in three volumes (Teubner, Leipzig, first edition, 1880-1; second edition, including the newly discovered Method, etc., 1910-15).

Of translations the following may be mentioned. The Aldine edition of 1558, 4to, contains the Latin translation by Commandinus of the Measurement of a Circle, On Spirals, Quadrature of the Parabola, On Conoids and Spheroids, The Sandreckoner. Isaac Barrow's version was contained in Opera Archimedis, Apollonii Pergoei conicorum libri, Theodosii Sphoerica, methodo novo illustrata et demonstrata (London, 1675). The first French version of the works was by Peyrard in two volumes (second edition, 1808). A valuable German translation, with notes, by E. Nizze, was published at Stralsund in 1824. There is a complete edition in modern notation by T. L. Heath (The Works of Archimedes, Cambridge, 1897, supplemented by The Method of Archimedes, Cambridge, 1912).

CHRONOLOGY.

(APPROXIMATE IN SOME CASES.)

B.C. 624-547 Thales 572-497 Pythagoras 500-428 Anaxagoras 470-400 / Hippocrates of Chios \ Hippias of Elis 470-380 Democritus 460-385 Theodorus of Cyrene 430-360 Archytas of Taras (Tarentum) 427-347 Plato 415-369 Theaetetus 408-355 Eudoxus of Cnidos / Leon fl. about 350 < Menaechmus | Dinostratus \ Theudius fl. 300 Euclid 310-230 Aristarchus of Samos 287-212 Archimedes 284-203 Eratosthenes 265-190 Apollonius of Perga